Table of Contents

Title Page v
INTRODUCTION vi
PREFACE x
CHAPTER ONE 1
    GETTING STARTED WITH BEEKEEPING .. 1
        Understanding Bees: Anatomy and Behavior 1
        What is beekeeping? 4
        History of beekeeping 5
        Benefits of Beekeeping: Environmental, Economical, and Personal 8
        Safety Precautions and Proper Protective Gear 10
CHAPTER TWO 13
    SETTING UP YOUR BEEKEEPING OPERATION 13
        Choosing the Right Hive: Langtroth, Top-Bar, or Warre 13
        Lang troth 15
        Parts of a Langstroth Hive 19
        Top-Bar 21

Warre 25
Location Selection 30
The Local Climate 31
Sunlight 31
Shade 33
The Right Bee Species 33
Flat and Dry Ground 34
Food Source 34
Beehive Position 35
Shelter 40
Water Source 41
Essential Equipment and Tools for Beekeeping 42

CHAPTER THREE 46

ACQUIRING BEES 46

Options for Obtaining Bees 46
Methods of acquiring Bee 47
Packages 48
Nucs 48
Swarms 49
Assessing Bee Health and Quality 50
Transporting Bees Safely to Their New Home 53

CHAPTER FOUR 57
  HIVE MANAGEMENT AND BEE CARE 57
    Bee Types 59
    The Life Cycle of Honeybees 61
    Hive Inspection 62
    Identifying and Managing Common Pests and Diseases 79
    Swarm Prevention and Management Strategies 82
  Honeybee Nutrition 85
    Macronutrients 86
    Micronutrients 87
    Basics of Bee Feeding 87
    Nutritional Sources 88
    Considerations for Climate 89
    Varieties of Bee Feeding Equipment 89
    Placement and Maintenance of Feeders 91
    Seasonal Feeding 91
    Observing and Modifying Feeding Practices 95
    Best Bee Feeding Practices 96
    Mitigating Disease and Pest Risks 97
CHAPTER FIVE 102

UNDERSTANDING BEE LIFE CYCLE AND COLONY DYNAMICS 102

 Roles Within the Hive..................................105

 Brood Development and Hive Population Dynamics 108

 Seasonal Considerations and Hive Management...110

CHAPTER SIX..115

HARVESTING HONEY AND OTHER HIVE PRODUCTS..115

 Assessing Honey Flow and Readiness for Harvest 115

 Harvesting Methods 118

 Processing Other Hive Products.................121

CONCLUSION...124

Title Page

The Backyard Beekeeper's Guide

A Complete Guide to Building Beehive, managing Colony, Harvesting Honey, and Turning Your Passion into Profit

By: James J. Bailey

INTRODUCTION

You jumped headlong into the project, full of enthusiasm to explore the world of backyard beekeeping. You felt ready for success since you had bees, a well-prepared hive, and all the necessary beekeeping equipment. But your hopes were short-lived when you saw a horrifying sight: a mass of dead bees scattered outside the hive entrance, indicating the end of your bee colony as a whole.

It's normal to experience shock, disappointment, and perhaps even a hint of guilt in these kinds of circumstances. This is actually more common than one might think among inexperienced beekeepers.

However, why does it occur?

Beginners frequently underestimate the difficulties and complications that come with beekeeping because they have an incorrect sense of readiness based on a superficial reading of publications or online tutorials. But the actual world is much more complex and demanding.

This is the key:
I tell you this not to scare you, but to emphasize how important it is to plan ahead when starting a beekeeping business. It is essential to conduct thorough research using books, courses, and instructional videos before to committing to this fulfilling endeavor.

Permit me to suggest a resource that had a significant impact on my experience as a beginning beekeeper and is still my first choice for prospective apiarists. Here I compile my experience: [The Backyard Beekeeper's guide A Complete Guide to Building Beehive, managing Colony, Harvesting Honey, and Turning Your Passion into Profit].

Why is this stage of preparation so important?
A failed hive can have far-reaching consequences for nearby bee populations in addition to causing

personal disappointment. Diseases or infestations that start in neglected hives can spread quickly and wipe out entire apiaries, affecting agricultural output as well as local ecosystems. The center of a beekeeper's business is an apiary, which acts as a haven where they care for their cherished colonies. These centers, which may be found anywhere from rural areas to suburban backyards, are essential to the maintenance of honeybee numbers and the promotion of pollination.

To summarize, ambitious backyard beekeepers can build a strong foundation by devoting themselves to instructional materials such as [The Backyard Beekeeper's guide A Complete Guide to Building Beehive, managing Colony, Harvesting Honey, and Turning Your Passion into Profit]. It not only broadens comprehension of the complex lives of bees, but it also gives people the know-how to avoid typical mistakes made by beginners.

Are you prepared to start your beekeeping journey? This fulfilling endeavor not only supports vital pollination efforts but also offers a rich and rewarding experience. Before diving in, it's essential to lay the groundwork by ensuring you have the appropriate habitat for nurturing honeybee colonies, selecting optimal hive locations, acquainting yourself with pertinent beekeeping guidelines and regulations, and acquiring other valuable insights to set you on the right path. Let's begin this journey together with a comprehensive understanding of the fundamentals to ensure your success and the well-being of your honeybee colonies. Let us embrace the wonder, the challenges, and the sheer magic of this ancient art, for in doing so, we not only cultivate honey but also harvest a bounty of wisdom, wonder, and wonderment.

Now that we have established this crucial foundation, let's move on to our next subject…

PREFACE

One of the most prevalent challenges confronting novice beekeepers is the lack of experience and expertise in adequately tending to a honeybee colony. This deficiency can manifest in various issues, including subpar hive management, insufficient bee nutrition, and a fundamental misunderstanding of bee behavior and requirements.

Bee, a tiny yet mighty creature that holds the strings to the melody of life: the honeybee. As an apiary enthusiast, picture this: A tranquil morning bathed in golden sunlight, the gentle hum of bees dancing amidst blossoms, and the air thick with the scent of nectar. This is the captivating scene that awaits those who dare to venture into the realm of beekeeping. This book is not merely a manual; it is a manifesto—a call to arms for the stewards of our planet, both seasoned apiarists and curious novices alike. Within these pages, you will find a treasure trove of knowledge, and experience the dance of bee as well as its behavior to the artistry of hive construction, from the sweet alchemy of honey extraction to the delicate level of hive management.

But beyond the practicalities lies a deeper resonance—For in the humble honeybee, we find a mirror reflecting our own fragility, resilience, and capacity for transformation. Through beekeeping, we not only nurture the health of our planet but also cultivate a profound sense of stewardship, mindfulness, and connection to the rhythms of life.

This point cannot be emphasized enough.

It is imperative to prioritize and invest substantially in your beekeeping education from the outset. Doing so will arm you with the requisite knowledge and skills to navigate the intricacies of beekeeping with confidence and proficiency as you embark on this fulfilling journey.

CHAPTER ONE
GETTING STARTED WITH BEEKEEPING

Understanding Bees: Anatomy and Behavior

Beekeeping, also referred to as apiculture, encompasses the human practice of nurturing honey bees. An individual engaged in this pursuit is commonly known as a beekeeper or apiarist. Their primary objective is to oversee honey bee colonies, either for harvesting honey and other hive-derived products like propolis, beeswax, and royal jelly, or for facilitating pollination in agricultural settings. Additionally, beekeepers may rear bees for commercial purposes, selling them to fellow apiarists.

Firstly, let's consider the anatomy of bees. These remarkable insects are comprised of several distinct parts, each serving a crucial function in their daily lives. At the forefront are their three main body segments: the head, thorax, and abdomen.

The head of a bee is a marvel of sensory perception and coordination. Here, we find their compound eyes, which provide them with a mosaic view of the world, allowing them to navigate with precision and detect the subtlest of movements. Additionally, bees possess antennae that serve as highly sensitive receptors, enabling them to perceive chemical cues and communicate with their fellow hive members through intricate dances.

Moving down the body, we encounter the thorax, where the powerhouse of the bee resides: the wings and muscles responsible for flight. Bees are marvels of aerodynamic efficiency, capable of navigating vast distances in search of nectar and pollen to sustain their colonies. Their wings beat with a rhythmic precision that is both mesmerizing and essential for their survival.

Finally, we come to the abdomen, where the magic of honey production takes place. Within this segment lies the honey stomach, a specialized chamber where bees store nectar collected from flowers. Through a process of enzymatic conversion and dehydration, this nectar is transformed into the golden elixir we know as honey—a vital source of energy and sustenance for the colony.

Now, let's turn our attention to the behavior of bees, which is equally fascinating and complex. Bees are highly social creatures, living in organized colonies with distinct roles and responsibilities. At the heart of the colony is the queen bee, the matriarch whose sole purpose is to lay eggs and ensure the survival of the hive. Surrounding her are the worker bees, sterile females tasked with myriad duties, from foraging and nursing to building and defending the hive. And let's not forget the drones, male bees whose primary function is to mate with the queen and perpetuate the colony's genetic lineage.

But perhaps most intriguing of all is the collective intelligence and communication exhibited by bees. Through a sophisticated system of pheromones, vibrations, and dances, bees are able to convey vital information about food sources, nest locations, and potential threats to their fellow colony members. This intricate language allows them to coordinate their efforts with uncanny precision, ensuring the prosperity and survival of the hive.

What is beekeeping?

Beekeeping is indeed a form of agricultural activity, known more broadly as apiculture. When stingless bees are employed in beekeeping, the terminology shifts to reflect the specific species of bee involved. The term "apiculture" typically pertains to beekeeping involving species from the Apis genus, which includes the commonly kept honey bee species such as Apis mellifera and Apis cerana.

Beekeeping can involve the management of both stinging and stingless bee species, each offering unique challenges and opportunities. Given the complexity of biological nomenclature and the challenge of recalling specific species names, it is generally acceptable to use the term "apiculture" when discussing any form of beekeeping, irrespective of the bee species involved. This umbrella term encompasses the art, science, and practical aspects of beekeeping, encompassing the lifecycle of honey bees and their adaptation to man-made structures in a comprehensive manner.

History of beekeeping
Beekeeping boasts ancient origins, evident through depictions in art and artifacts dating back over 10,000 years. In antiquity, early beekeepers utilized pottery vessels as rudimentary hives to accommodate honey bee colonies. Over time, advancements led to the development of more sophisticated hive structures and the use of smoke as a tool in beekeeping practices. Eventually, the evolution of beekeeping culminated in the storage of honey in jars, marking a significant milestone in the history of apiculture.

During the 18th century, European interest in beekeeping surged, prompting extensive studies into the life cycle and biology of honey bees. The aim was to gain insights into honey bee colony organization. This deeper understanding led to a pivotal discovery known as 'bee space', which revolutionized beekeeping practices. The concept of bee space facilitated the development of movable frames within beehives, laying the foundation for the artificial beehives widely utilized in modern beekeeping.

Throughout its history, beekeeping has employed various hive designs, including skeps, pottery vessels, log hives, top bar hives, Warré hives, and the Langstroth beehive. Among these, the Warré, Langstroth, and Top Bar hives have emerged as the most popular choices in contemporary beekeeping practices. Later in this article, we will delve into these hive designs in greater detail.

In recent years, a novel adaptation of the Langstroth beehive, known as the Flow hive, has gained popularity among beekeepers. The Flow hive incorporates proprietary technologies that enable the harvesting of honey without the need to remove beehive frames from the hive. Specifically designed frames tailored for the Flow hive facilitate this innovative harvesting method.

While Flow hives offer unparalleled convenience in honey extraction, they present limitations in harvesting other bee products typically obtained by beekeepers. Notably, relying solely on a Flow hive setup may result in minimal beeswax yield at the end of each beekeeping season. However, Flow hives yield exceptionally pure honey with minimal impurities, and they significantly streamline the beekeeping process, saving beekeepers valuable time and eliminating the need for a separate honey extractor, thereby reducing costs.

Until recent times, beekeepers relied heavily on smoke to deter honey bees from stinging during hive inspections. Prior to the widespread use of beekeeping veils, beekeeping was a relatively hands-off practice, with hive visits primarily focused on assessing honey storage levels and harvesting. However, modern beekeeping demands more frequent monitoring and management of honey bee colonies and their hives.

Benefits of Beekeeping: Environmental, Economical, and Personal

Bees play a pivotal role in pollinating flowering plants, including many of the fruits, vegetables, and nuts that comprise a significant portion of our diets. By maintaining healthy bee populations, beekeepers contribute to the biodiversity of ecosystems and help ensure the continued abundance of agricultural crops. In essence, beekeeping is a form of environmental stewardship, promoting the health and vitality of our planet's ecosystems.

Beekeeping can also be a lucrative enterprise, providing beekeepers with a source of income through the sale of honey, beeswax, pollen, and other hive products. Additionally, beekeeping can bolster local economies by supporting agricultural industries that rely on pollination services provided by bees. From small-scale hobbyists to commercial apiarists, beekeepers play a vital role in sustaining rural livelihoods and fostering economic resilience.

For many enthusiasts, beekeeping is not just a hobby but a deeply fulfilling passion that brings joy, satisfaction, and a sense of connection to nature. The act of tending to bees, harvesting honey, and witnessing the intricate workings of the hive can be profoundly rewarding experiences that foster a deeper appreciation for the natural world. Moreover, beekeeping offers opportunities for personal growth and learning, as beekeepers continually refine their skills, expand their knowledge, and cultivate a deeper understanding of bees and their behaviors.

Safety Precautions and Proper Protective Gear

It is essential to recognize that beekeeping, while immensely rewarding, also comes with inherent risks, particularly in terms of bee stings. Therefore, adopting appropriate safety measures is paramount to ensure the well-being of both beekeepers and their bees.

The cornerstone of beekeeping safety is the use of proper protective gear. This includes a beekeeping suit, veil, gloves, and boots—all designed to provide a physical barrier between beekeepers and their buzzing charges. The beekeeping suit, typically made of lightweight, breathable fabric, covers the entire body from head to toe, offering protection against bee stings. The veil, attached to the suit, shields the face and neck from bee encounters while still allowing for clear vision and airflow. Additionally, thick gloves shield the hands and wrists from stings, while sturdy boots provide protection for the feet.

In addition to wearing protective gear, beekeepers should also exercise caution and mindfulness when working with bees. This includes approaching hives calmly and quietly, avoiding sudden movements or loud noises that may agitate the bees. Furthermore, beekeepers should be vigilant in observing bee behavior, as changes in buzzing intensity or defensive posturing may indicate that the bees feel threatened and are preparing to sting.

Beyond personal protective equipment, beekeepers should also be equipped with essential first aid supplies to treat bee stings promptly. This includes antihistamines, topical ointments, and sting relief products to alleviate discomfort and reduce the risk of allergic reactions.

Lastly, beekeepers should be mindful of their own health and well-being, particularly when working in hot weather or during extended hive inspections. Staying hydrated, taking breaks as needed, and listening to your body's cues are essential practices to prevent heat exhaustion and fatigue.

CHAPTER TWO
SETTING UP YOUR BEEKEEPING OPERATION

Choosing the Right Hive: Langtroth, Top-Bar, or Warre

When it comes to beekeeping, selecting the appropriate hive is a pivotal decision that can significantly impact the health and productivity of your colony. While there are various hive designs available, three of the most common options are the Langstroth, Top-Bar, and Warre hives.

Each type of hive, whether it's a Warre or Langstroth hive, presents its own distinct advantages and disadvantages. While many beekeepers prefer the Langstroth hive for its user-friendly design and higher honey production, the Warre hive is favored for its ability to replicate natural beekeeping conditions, potentially promoting bee health and well-being.

One of the primary differences between Warre's and Langstroth's hives lies in their structure. A Warre hive comprises a series of uniform, smaller boxes, whereas a Langstroth hive can incorporate various sizes of super boxes: deeps/brood, mediums, or shallows, although uniform-sized boxes are also an option.

In a Warre hive, boxes are added from the bottom, necessitating some initial heavy lifting during the nectar flow period. However, since Warre hives are less frequently manipulated, heavy lifting is not as regular a task.

Conversely, Langstroth supers are added above the brood box. Accessing the brood box for inspection requires removing all additional boxes, which can be cumbersome.

Ultimately, the choice between a Warre hive and a Langstroth hive hinges largely on personal preference. Bees adapt to their environment regardless of hive type. Neither guarantees bee health or success; that responsibility lies with the beekeeper.

Lang troth

The Langstroth hive, which is perhaps the most widely used and recognized hive design among beekeepers. Named after its inventor, Reverend Lorenzo Lorraine Langstroth,

The Langstroth Hive stands as a cornerstone of modern beekeeping, widely utilized and celebrated for its efficiency. Developed by Lorenzo Langstroth, an American beekeeper, in the mid-19th century, this hive owes its innovation to Langstroth's discovery of "bee space." This concept refers to the gap bees naturally maintain between adjacent honeycombs.

This hive features vertically stacked boxes with removable frames. The modular design allows for easy inspection, manipulation, and management of the hive, making it an ideal choice for both beginners and experienced beekeepers alike. Additionally, Langstroth hives offer flexibility in hive expansion and honey extraction, as well as compatibility with standard beekeeping equipment and accessories.

This groundbreaking design enables beekeepers to manage their colonies effectively without disrupting the bees' habitat. Langstroth's invention revolutionized the beekeeping industry, offering unparalleled convenience and productivity. Today, the Langstroth Hive remains a favored choice among beekeepers worldwide, testament to its enduring impact and utility in modern apiculture.

The Langstroth hive streamlines hive management and honey harvesting with its use of frames equipped with pre-built comb foundations, offering beekeepers enhanced ease and efficiency.

Similar to the Warre hive, the Langstroth hive comprises multiple stacked boxes. As the colony grows, new boxes are added to the top of the hive (supering), accommodating its expansion. Additionally, the Langstroth hive includes a bottom board to facilitate waste and debris management, along with an entrance positioned toward the bottom of the hive for easy access.

The Langstroth hive stands out for its versatility and adaptability, offering beekeepers greater flexibility and control over their hives, particularly during the nectar-flow season. This hive type is particularly well-suited for commercial beekeeping endeavors, as it facilitates larger honey yields and provides convenient access to brood frames for inspections.

Effective management of a Langstroth hive entails regular inspections aimed at monitoring the colony's health and productivity. Beekeepers frequently engage in activities such as rearranging frames, managing swarm behavior, and replacing queen bees as needed. Moreover, Langstroth hives can be easily expanded or condensed to accommodate the growth of the colony, making them an ideal choice for scalable beekeeping operations.

Maintaining a Langstroth hive can indeed be more labor-intensive compared to a Warre hive, as it involves handling multiple heavy hive boxes. Harvesting honey from a Langstroth hive entails the removal of individual frames, uncapping the cells, and utilizing a honey extractor to extract the honey from the wax. While this process may be time-consuming, it yields a cleaner honey product and allows for the reuse of the wax combs.

The Langstroth hive stands as the most widely employed type of beehive globally, renowned for its exceptional honey production capabilities. Its innovative design incorporates removable frames that guide bees to construct their comb precisely where the beekeeper requires it, facilitating efficient colony management and honey extraction while minimizing disruption to the bees.

Moreover, the Langstroth hive's vertical expansion contributes to its heightened honey output. As the colony expands, additional boxes, known as supers, are stacked atop the existing ones. This vertical growth pattern affords ample space for honey storage, thereby fostering increased honey production.

Parts of a Langstroth Hive

Outer/Telescoping Cover: Resembling a roof on a house, this cover shields the entire hive from rain, ensuring its dryness.

Inner Cover: Positioned between the top hive box and the outer cover, the inner cover furnishes insulation and prevents frames from adhering to the outer cover. Additionally, it can be used in conjunction with a bee escape during honey harvesting.

Shallow/Honey Super: These shallow supers are predominantly utilized for honey production due to their common size.

Queen Excluder: An optional component, the queen excluder permits only worker bees to traverse, effectively segregating the queen and drones from the honey collection area. It serves to prevent the queen from laying eggs in the honey supers, although not all beekeepers opt to use it.

Frames: Removable frames, crafted from wood or plastic, slot into the hive boxes. Available in various sizes to accommodate the three different super sizes, these frames house honeycomb constructed by the bees, often with the aid of beeswax or plastic foundation.

Foundation: Sheets of beeswax or plastic foundation are commonly utilized by beekeepers as a guide within the frames, encouraging the bees to construct straight comb inside.

Brood Chamber (Also referred to as a deep super or brood box): Housing larger frames than the shallow supers, the brood chamber serves as the site where the queen lays eggs for the forthcoming generation of bees. Acting as a maternity ward, nurse bees tend to the young within this chamber.

Bottom Board: Serving as the hive's base, the bottom board is available in options such as a solid bottom or a screened bottom, providing ventilation.

A Langstroth hive can accommodate any combination of the three sizes of super boxes: deeps/brood, mediums, or shallows.

Top-Bar

The Top-Bar hive is a more traditional and minimalist alternative to the Langstroth hive. Despite common misconceptions, top bar beekeeping has roots dating back centuries, reflecting a longstanding practice rather than a recent innovation. Historically, beekeepers utilized rudimentary methods by placing strips of wood over vacant cavities, encouraging bees to construct comb from beneath.

In contemporary beekeeping, two primary styles of horizontal top bar hives prevail: The Kenyan top bar hive and the Tanzanian top bar hive. The Kenyan variant features angled, elongated sides, yielding a trapezoidal hive shape, whereas the Tanzanian counterpart boasts straight sides with 90-degree angles, akin to a lengthy rectangular box.

The Kenyan top bar hive originated in 1971 through the ingenuity of two Canadian beekeepers seeking an economical and uncomplicated alternative to Langstroth hives. This initiative stemmed from a development program underway in Kenya at the time. The original design featured a hollow, trapezoidal cavity covered by a row of top bars, which fully encased the internal hive space. Bees commenced comb construction downward from these top bars, maneuvering horizontally through the hive in either left or right directions.

In the Kenyan top bar hive design, the active hive space is bordered on both sides by a movable wall, known as a "divider board," enabling beekeepers to adjust the hive space according to seasonal requirements. This feature offers enhanced flexibility, allowing for expansion or contraction of the hive as necessary. We favor this style of top bar hive over the Tanzanian variant due to its superior comb stability, particularly in hot summer conditions.

In a top bar hive, the width of the top bars is sufficient to accommodate bee space on both sides of the comb. Numerous top bars incorporate a central guide to facilitate the bees in building their comb along the center of the top bar, rather than off to one side. These guides may be integrated into the top bar design as a wedge or created by adhering popsicle sticks along the center of the bar.

Despite the presence of a guide, bees may occasionally construct comb off-center on top bars, a phenomenon known as cross-combing. While cross-combing can occur in any hive, it is of particular concern in top bar hives. To mitigate cross-combing, it is advisable to conduct frequent hive inspections, ideally every 1-2 weeks following the initial installation of bees.

During inspections, any comb that deviates from a straight formation can be gently manipulated into shape or trimmed off and reattached using zip ties. Once several straight combs have been established, alternating between built comb and empty bars can aid in guiding the bees and minimizing the occurrence of cross-combing.

In contrast to the stacked boxes of the Langstroth, the Top-Bar hive consists of a single horizontal cavity with bars across the top, onto which bees build their comb. This design mimics the natural architecture of wild beehives and offers several advantages, including ease of construction, minimal equipment requirements, and reduced disturbance to the bees during inspections.

However, Top-Bar hives may have lower honey yields compared to Langstroth hives and require more frequent and careful management to prevent comb collapse and maintain hive health.

Warre

The Warre hive is named after its creator, French beekeeper Emile Warre. The Warre Hive, often referred to as the "people's hive," was conceived by Emile Warre, a notable French beekeeper. His objective was to devise a straightforward yet efficient system that emulated the natural nesting behaviors of honeybees. The Warre Hive is designed to facilitate the construction of honeycombs by bees in a manner akin to their natural habitat, a feature that has propelled its popularity among beekeepers seeking a more organic approach to beekeeping.

The Warre hive is characterized by its smaller boxes and utilization of top bars instead of frames, enabling bees to construct their comb in a natural manner. This design promotes a hands-off approach to beekeeping and is cost-effective to either construct or purchase.

Consisting of multiple stacked boxes with a quilt box positioned on top, the hive is engineered to offer insulation and ventilation to the colony. The hive entrance is situated at the base, while additional boxes are added beneath the existing structure as the colony expands, a process known as nadiring.

Similar to the Top-Bar hive, the Warre hive features a horizontal design with stacked boxes, each containing top bars for comb construction. However, unlike the Top-Bar hive, the Warre hive utilizes a vertical expansion method known as nadiring, in which additional boxes are added to the bottom of the hive as the colony grows. This approach aims to mimic the natural swarming behavior of bees and encourages colony expansion while minimizing disturbance. Warre hives are prized for their simplicity, sustainability, and bee-centric design, making them a favorite among natural beekeeping enthusiasts.

Beekeepers who opt for Warre hives typically embrace a hands-off management approach, allowing the bees to operate with minimal interference. While this method doesn't ensure immunity from mites, it increases the likelihood of healthy and secure bee colonies.

Regarding maintenance, harvesting honey from a Warre hive tends to be less labor-intensive compared to a Langstroth hive. At the close of the season, beekeepers simply remove the boxes containing honey and replace them with empty ones, enabling the bees to proceed with comb-building unhindered.

In production of honey, the Warre hive typically produces less honey in comparison to other hive types, such as the Langstroth hive.

Given that bees in a Warre hive have the freedom to place brood anywhere they choose, achieving entire boxes of pure honey poses a greater challenge. Additionally, the smaller size of the boxes in a Warre hive results in a reduced volume of honey stored within, further impacting overall honey production.

The Warré Hive offers a notable advantage in its design, fostering a natural environment for bee operation akin to their natural nests. This hands-free management approach requires minimal intervention from beekeepers compared to other hive types.

However, the Warré Hive does present certain drawbacks. Inspecting individual bars of honeycomb is challenging, and the absence of frames makes comb more susceptible to damage.

Moreover, the risk of cross-comb formation poses difficulties in hive inspection, potentially rendering the colony difficult to assess.

Harvesting honey from a Warré hive may also pose challenges. The process involves cutting, crushing, and straining the comb, which is less efficient, particularly for beekeepers managing multiple hives.

Location Selection

Beekeeping commences with a crucial initial step — selecting the optimal location for your beehives. Amidst a myriad of factors ranging from soil composition to wind patterns, navigating this decision ensures your bees are situated in an environment primed for flourishing. Feeling uncertain about where to initiate your beekeeping endeavor? Fear not! This guide, as promised, is poised to navigate you through the indispensable components of choosing the perfect locale for your hives.

Beekeeping transcends merely establishing a hive and anticipating honey production. It necessitates meticulous planning, and the selection of your beehive's location holds paramount importance. A thoughtfully chosen site ensures:

Healthy Bees: Opting for a locale abundant in forage and water sources promotes the overall well-being of the bees.

High Productivity: Bees thrive in environments rich with diverse flora, resulting in increased honey production.

Protection: A well-sited hive offers safeguarding against predators and adverse weather conditions, preserving the colony's stability and productivity.

The Local Climate

Understanding the local climate is paramount as you embark on your beekeeping journey. Each bee species harbors distinct requirements concerning climate conditions. For instance, residing in regions marked by scorching summers necessitates placing your beehive in a shaded area. Conversely, in locales characterized by frigid winters, ensuring your beehive basks in ample sunlight becomes imperative.

Sunlight

Sunlight plays a pivotal role in the health and productivity of your bee colony. Bees are cold-blooded creatures, relying on external heat sources to regulate their body temperature and maintain hive conditions. Therefore, it's essential to position your hive in a location that receives ample sunlight throughout the day, particularly in the morning when bees are most active. Ideally, aim for a spot with exposure to the east or southeast, allowing the hive to bask in the warmth of the morning sun and kickstart the day's foraging activities.

Sunlight plays a crucial role in beekeeping as it aids bees in regulating their body temperature. Ensuring adequate exposure to sunlight is essential for bees to maintain warmth during winter and stay cool amid summer heat. Moreover, sufficient ventilation within the beehive is imperative. Bees generate substantial heat during their activities, necessitating effective mechanisms to dissipate this heat. Inadequate ventilation can lead to overheating, endangering the colony's survival.

Shade

Providing shade for bees is crucial, especially during hot summer months. Excessive heat can be detrimental to bees, leading to heat stress and potential colony loss. Adequate shade helps bees regulate their internal temperature, preventing overheating and ensuring their well-being. Additionally, shade complements the need for proper ventilation, allowing bees to dissipate heat effectively and maintain a comfortable hive environment.

The Right Bee Species

Selecting the appropriate bee species for your region is essential for successful beekeeping. Different bee species have specific requirements and adaptations to various climates and environments. Conduct thorough research to determine which bee species are best suited to your area's climate and conditions. Consider factors such as temperature, humidity, and forage availability when choosing the right bee species for your beekeeping venture. By selecting the most suitable bee species, you can ensure the health and productivity of your bee colonies.

Flat and Dry Ground

When selecting a location for your beehive, prioritize flat and dry ground. Opting for level ground ensures stability for the hive, facilitating easy movement for the bees inside. Avoid uneven terrain, as it can impede bee activity within the hive. Additionally, steer clear of areas prone to moisture or mold issues, as these conditions can be detrimental to bee health and comb construction. Excess moisture may also pose a risk of hive instability, potentially causing the hive to tip over. Therefore, aim for a dry spot free from moisture-related challenges to promote optimal hive conditions.

Food Source

Ensure there is a readily available food source nearby for your bees. Flowers are essential for honey production, so it's vital to have an abundance of blooming plants in close proximity to your hive. Consider planting a diverse garden with flowers that bloom at different times throughout the year, providing a continuous supply of nectar and pollen for the bees. Properly maintain the garden by watering the plants and ensuring their health to sustain the bees' food source consistently. By cultivating a garden with suitable flowers, you can support the bees in maintaining a steady food supply essential for their well-being and honey production.

Beehive Position
Positioning your beehive is a critical aspect of successful beekeeping. A well-placed hive not only promotes the safety and health of your bees but also enhances honey production. Careful consideration of the hive's location ensures the productivity and contentment of your bees over time.

Bees rely on the sun for navigation, so it's essential to position the hive in an area that receives ample sunlight. Avoid placing the hive in shady spots, as this can disorient the bees and make it challenging for them to return to the hive. Ensure there are no obstructions such as trees or buildings blocking the sunlight, allowing the bees to navigate easily.

Additionally, consider the surrounding landscape when positioning your beehive. Choose a location with minimal wind exposure to provide protection from harsh weather conditions. Avoid areas prone to flooding or excessive moisture, as these conditions can negatively impact hive health.

Taking the time to select the right location for your beehive is crucial for the long-term success of your beekeeping venture. By providing your bees with a suitable environment, you can support their well-being and maximize honey production for years to come.

Positioning your beehive with the entrance facing southwest offers several benefits for hive health and productivity. Placing the hive in a location that receives early morning sun and afternoon shade ensures optimal working conditions for the bees, maintaining an ideal temperature within the hive throughout the day. Additionally, this orientation facilitates the bees' navigation, providing them with the best chance of returning safely to the hive after foraging.

In addition to sunlight considerations, it's essential to provide a windbreak to protect the hive from strong winds. A fence, hedge, or group of trees can serve as effective wind barriers, shielding the hive and minimizing the risk of damage caused by turbulent weather conditions.

By carefully positioning your beehive to optimize sunlight exposure, temperature regulation, and wind protection, you create a conducive environment for bee activity and overall hive well-being. This thoughtful approach enhances the success of your beekeeping endeavors and promotes thriving bee colonies.

Elevating your hive on a stand serves multiple purposes in ensuring hive safety and longevity. By keeping the hive off the ground, you protect it from potential flooding and other ground-related hazards. Optimal elevation for a hive stand is approximately 18 inches, providing adequate clearance while facilitating hive access for maintenance and inspection.

In addition to raising the hive, implementing a windbreak is essential for safeguarding against strong winds. Whether it's a fence, hedge, or clustered trees, a windbreak shields the hive, mitigating the risk of damage caused by turbulent weather conditions.

By combining the elevation of the hive on a stand with the installation of a windbreak, you create a protective environment that enhances hive resilience and longevity. These measures contribute to the overall health and productivity of your beekeeping operation, ensuring the well-being of your bee colonies in various environmental conditions.

Maintaining adequate spacing between beehives is crucial for ensuring the comfort and well-being of your bee colonies. In areas with numerous hives, sufficient space between them prevents overcrowding, allowing bees to move freely without feeling confined.

Moreover, providing ample clearance in front of each hive ensures unobstructed flight paths for incoming and outgoing bees. To achieve optimal spacing, it's recommended to maintain a minimum distance of at least 5 feet in all directions surrounding each hive. This ensures sufficient room for bees to maneuver and facilitates efficient hive management practices.

Regular access to your beehive is essential for monitoring the well-being of your bees and ensuring their optimal care. Placing your hive in an easily accessible location simplifies hive inspections and maintenance tasks.

Additionally, prioritizing the comfort of your bees is crucial for their overall health. Choosing a spot with adequate airflow helps regulate the hive's temperature, ensuring a cool and comfortable environment for the bees. By selecting a location with proper ventilation, you can promote a thriving bee colony and facilitate effective hive management practices.

Shelter

Shelter is a crucial consideration for protecting your hive from the elements and ensuring the comfort and well-being of your bees. Choose a location that offers natural windbreaks, such as hedges, trees, or buildings, to shield the hive from strong winds and inclement weather. Additionally, consider the orientation of the hive entrance relative to prevailing winds to minimize drafts and maintain hive temperature and humidity levels. A sheltered location not only provides a more stable environment for your bees but also reduces stress and energy expenditure, allowing them to focus their efforts on foraging and hive maintenance.

Water Source

Ensure there is a reliable water source in close proximity to your beehive. Bees require water for hydration, and they will travel considerable distances to find suitable water sources. Without access to water nearby, bees may suffer from dehydration and fail to thrive. Ideally, locate your beehive near a pond, lake, river, or other natural water bodies to provide the bees with easy access to water. Additionally, consider placing a shallow dish of water near the hive to supplement their water supply. However, ensure that the water dish is shallow to prevent bees from drowning. Providing a readily available water source near your beehive is crucial for the bees' well-being and overall hive productivity.

Bees require water for various essential functions, including cooling the hive, diluting honey for consumption, and rearing brood. Therefore, it's essential to locate your hive near a clean, accessible water source, such as a pond, stream, or birdbath. Ensure that the water source is shallow and has a secure landing area for bees to safely collect water without risk of drowning. By providing your bees with easy access to water, you not only support their health and vitality but also reduce the likelihood of them seeking out alternative—and potentially less desirable—water sources, such as swimming pools or bird feeders.

Essential Equipment and Tools for Beekeeping
Starting a hobbyist beekeeping journey is an exciting endeavor, but it's essential to have the right equipment to ensure the success of your beekeeping venture. While the options may seem overwhelming at first, understanding the essential beekeeping equipment can help you get started on the right foot.

First and foremost, you'll need to acquire the bees themselves, whether through purchasing a package of bees or obtaining a nucleus colony (nuc). Once you have your bees, it's time to invest in the necessary equipment and tools to support your hive.

Essentially, every beekeeper needs a hive—the structure where bees will live and produce honey. The type of hive you choose, whether it's a Langstroth, Top-Bar, or Warre hive, will depend on your preferences and beekeeping goals.

In addition to the hive itself, several essential tools are required to properly manage and maintain your bee colony:

1. Hive Tool: This versatile tool is a beekeeper's best friend, used for prying apart hive components, scraping propolis and wax, and manipulating frames during inspections.

2. Smoker: A smoker is essential for calming bees during hive inspections. By puffing cool smoke into the hive, beekeepers can disrupt the bees' communication signals and reduce the likelihood of defensive behavior.

3. Bee Brush: A gentle bee brush is used to gently sweep bees off frames and hive components during inspections, allowing beekeepers to access and inspect the hive without squashing bees or causing undue stress.

4. Protective Gear: Beekeeping can be a sting-prone activity, so proper protective gear is essential for staying safe. This includes a beekeeping suit, veil, gloves, and boots, all of which provide a physical barrier between beekeepers and their buzzing charges.

5. Feeder: Depending on the season and hive conditions, beekeepers may need to supplement their bees' diet with sugar syrup or other supplemental feed. A variety of feeders are available, including entrance feeders, top feeders, and frame feeders, each offering different advantages depending on your needs.

6. Queen Excluder: This specialized device is used to prevent the queen bee from laying eggs in honey supers, ensuring that harvested honey remains free of brood.

7. Honey Extraction Equipment: If you plan to harvest honey from your hive, you'll need equipment for extracting, filtering, and bottling honey. This includes a honey extractor, strainers, settling tanks, and jars for storing and packaging honey.

8. Beekeeping Books and Resources: Lastly, investing in educational materials, such as beekeeping books, guides, and online resources, is essential for expanding your knowledge and honing your beekeeping skills.

By acquiring these essential equipment and tools, you'll be well-equipped to embark on your beekeeping journey with confidence and success. Remember to invest in quality equipment, prioritize safety, and continually educate yourself to become a skilled and responsible beekeeper.

CHAPTER THREE
ACQUIRING BEES

Options for Obtaining Bees

Beekeeping offers a rewarding and captivating hobby that not only aids in supporting the declining bee population but also provides a fresh supply of honey straight from your own backyard!

When starting your beekeeping journey, one of the initial steps is to source and acquire bees.

Selecting the appropriate bee species is paramount in beekeeping, ensuring they are well-suited to your local environment.

Various bee breeds exhibit distinct characteristics and adaptability to environmental conditions, necessitating careful consideration to ensure the bees thrive in your region.

Here are pivotal factors to weigh when choosing the ideal bee species for your locale:

Hardiness: Opt for bee breeds renowned for their resilience against prevalent diseases and pests in your area, promoting hive health and longevity.

Climate: Assess the temperature fluctuations and weather patterns characteristic of your region. Certain bee varieties excel in colder winters, while others flourish in warmer climates, aligning with your locale's climatic conditions.

Forage Availability: Evaluate the abundance of nectar and pollen-producing flora in your vicinity. Bees rely on a diverse range of forage to maintain robust health and productivity, emphasizing the importance of ample foraging resources.

Methods of acquiring Bee

Acquiring bees for your hive involves several methods, each with its unique advantages and considerations. Knowing the disparities between packages, NUCs (nucleus colonies), and swarms aids in making an informed decision tailored to your experience level and preferences. Let's examine each:

Packages

Bee packages—a popular choice for many beginner beekeepers. A package of bees typically consists of a group of worker bees and a mated queen bee housed in a ventilated box. These bees are usually sourced from commercial bee breeders and are shipped to beekeepers in the spring when bee populations are at their peak. Packages offer a convenient and relatively inexpensive way to start a new colony, allowing beekeepers to introduce bees to their hive and establish a thriving colony from scratch.

Nucs

The nucleus colonies, commonly referred to as Nucs. A nuc is a small, established colony of bees that typically contains several frames of brood, worker bees, a mated queen, and stores of pollen and honey. Nucs are often sold by experienced beekeepers or bee breeders and provide a more advanced starting point for beekeepers looking to jumpstart their hive. Nucs offer several advantages over packages, including a head start in colony development, increased likelihood of survival, and reduced risk of queen failure.

Swarms

Swarms—a natural phenomenon that occurs when a colony of bees outgrows its hive and the queen bee leads a portion of the colony in search of a new home. Swarms are typically captured by experienced beekeepers or bee removal specialists and transferred to a hive for rehoming. While capturing swarms can be unpredictable and requires specialized skills and equipment, it offers an opportunity to acquire bees for free and provides a valuable service in helping to control feral bee populations and prevent conflicts with humans.

Indeed, each method of acquiring bees has its unique considerations, necessitating an evaluation of your comfort level, budget, and desired involvement level before making a choice. Having gained insight into the different types of bees, how to identify reputable bee breeders and suppliers, and the various methods of acquiring bees, you are now better equipped to embark on your beekeeping journey. With careful planning and informed decision-making, you'll be well-prepared to nurture and sustain a thriving bee colony in your backyard.

Assessing Bee Health and Quality

Assessing bee health is a fundamental responsibility of every beekeeper, as it directly impacts the productivity, vitality, and survival of the colony. Here are some key factors to consider when evaluating the health and quality of your bees:

1. Population Size: One of the first indicators of bee health is the size and activity level of the colony. A healthy hive will have a robust population of worker bees, bustling with activity as bees come and go on foraging flights. Conversely, a declining or weak colony may exhibit reduced activity at the hive entrance and fewer bees present inside the hive.

2. Brood Patterns: Inspecting the brood frames for healthy brood patterns is essential for assessing the reproductive health of the colony. Healthy brood frames should contain tightly capped cells of developing brood in various stages—eggs, larvae, and capped pupae—indicating a thriving queen and successful brood rearing.

3. Queen Performance: The performance of the queen bee is critical to the overall health and productivity of the colony. During hive inspections, beekeepers should observe the presence of eggs, larvae, and capped brood in sufficient quantities, as well as the presence of a healthy and active queen bee. Signs of queen failure, such as a lack of brood or the presence of drone-laying workers, may indicate issues with queen health or mating.

4. Disease and Pest Management: Regular monitoring for signs of diseases and pests is essential for maintaining bee health and preventing outbreaks. Common diseases and pests include varroa mites, Nosema, American foulbrood, and chalkbrood. Beekeepers should familiarize themselves with the symptoms and signs of these ailments and implement appropriate management strategies, such as integrated pest management (IPM) practices and medication treatments when necessary.

5. Foraging Behavior: Observing the foraging behavior and activity of bees can provide valuable insights into hive health and environmental conditions. Healthy colonies will exhibit strong foraging activity, with bees returning to the hive laden with pollen and nectar. Changes in foraging behavior, such as reduced activity or orientation flights, may indicate issues with hive health, nutrition, or environmental stressors.

6. Hive Environment: Lastly, maintaining a clean, dry, and well-ventilated hive environment is essential for promoting bee health and preventing disease. Regular hive maintenance, including cleaning and replacing old comb, managing hive ventilation, and providing adequate space for colony expansion, can help create optimal conditions for bee health and productivity.

Transporting Bees Safely to Their New Home
Transporting bees requires careful planning, preparation, and attention to detail to ensure the well-being of both the bees and the beekeeper. Here are some essential steps to follow when transporting bees to their new hive:

1. Prepare the Hive: Before transporting bees, ensure that the hive is properly assembled, secure, and equipped with all necessary components, including frames, foundation wax, and entrance reducers. It's also a good idea to close and secure any openings or entrances to prevent bees from escaping during transit.

2. Choose the Right Time: Selecting the right time of day for transporting bees is crucial to minimize stress and maximize safety. Ideally, transport bees during the early morning or late evening when temperatures are cooler, and bees are less active. Avoid transporting bees during the hottest part of the day to prevent overheating and stress.

3. Secure the Hive: Once the hive is prepared, securely fasten it to a stable base or platform inside the vehicle to prevent shifting or tipping during transit. Use straps or bungee cords to secure the hive firmly in place, ensuring that it remains stable and upright throughout the journey.

4. Provide Adequate Ventilation: Proper ventilation is essential to ensure the comfort and well-being of the bees during transport. Ensure that the hive is adequately ventilated to prevent overheating and suffocation, particularly during hot weather. If transporting bees in a closed vehicle, crack windows or provide additional ventilation to promote airflow.

5. Handle Bees Carefully: When handling bees during transport, use caution and gentle movements to minimize stress and agitation.

Avoid jostling or shaking the hive, as this can disturb the bees and provoke defensive behavior. If necessary, use a smoker to calm the bees before and during transport.

6. Monitor Temperature: Be mindful of temperature fluctuations during transport, as extreme heat or cold can pose risks to bee health. Take precautions to protect bees from overheating or exposure to cold temperatures by providing shade, insulation, or additional ventilation as needed.

7. Plan Your Route: Before setting out, plan your route carefully to minimize travel time and avoid rough roads or bumpy terrain that could jostle the hive. Allow for sufficient rest stops to check on the bees, monitor hive conditions, and address any issues that may arise during transit.

8. Be Prepared: Finally, be prepared for unexpected challenges or emergencies that may arise during transport. Carry essential beekeeping supplies, such as a smoker, protective gear, first aid kit, and tools, to address any issues that may arise en route.

By following these guidelines and taking proper precautions, beekeepers can safely transport bees to their new home, ensuring a smooth transition and setting the stage for a successful beekeeping endeavor.

CHAPTER FOUR
HIVE MANAGEMENT AND BEE CARE

Beekeeping is both captivating and fulfilling, but it necessitates a deep understanding and adeptness. Among the vital aspects of beekeeping lies beehive management. Proper management practices are pivotal in maintaining the health and productivity of your bee colony, while negligence can precipitate a myriad of challenges.

Beehive management encompasses a spectrum of tasks, including routine hive inspections, sustenance provision, and pest and disease control. Regular inspections are imperative to ascertain the hive's well-being and functionality. During these checks, beekeepers scrutinize for signs of diseases, pests, and other potential issues that might affect the bees. They also evaluate honey reserves and ensure the queen's egg-laying activity and the health of the brood. Provision of sustenance is another critical facet of beehive management. Bees rely on a constant supply of nectar and pollen for sustenance and honey production. In the event of scarce food sources, beekeepers may need to supplement the bees' diet with sugar water or pollen patties.

In addition, effectively managing your honeybee hive is fundamental in the practice of beekeeping. This encompasses essential tasks, including comprehending the lifecycle of honeybees, conducting routine hive inspections, implementing maintenance procedures, and addressing common bee diseases and pests. Additionally, it involves the rewarding endeavor of harvesting and extracting honey. Effective pest and disease management are paramount for hive health. Beekeepers must be vigilant in monitoring pests like mites and beetles, along with diseases such as foulbrood. By adopting proactive measures to prevent and address these challenges, beekeepers can safeguard the vitality and productivity of their hives.

Comprehending the lifecycle of honeybees is paramount for effective hive management. The lifecycle commences with the queen bee, responsible for laying eggs within the comb.

Bee Types

Subsequently, these eggs undergo hatching, transforming into larvae, which receive care and sustenance from worker bees until they advance into the pupal stage, while adult bees emerge from the pupae, assuming diverse roles within the hive community. This comprehension of bee development empowers beekeepers to discern the distinct stages and make well-informed decisions concerning hive management.

Within a honeybee hive, a multitude of bees undertake unique roles crucial for its overall operation. Comprehending these diverse responsibilities is pivotal to grasp the intricacy of the hive's social framework. Here are the key types of honeybees you'll encounter:

Bee Type	Description
Queen Bee	The queen bee holds the central role within the hive as its primary reproducer. Her duties encompass laying eggs to ensure the hive's population sustenance and fostering a cohesive social order among the bees. Additionally, her pheromones function as a vital means of communication within the hive community.
Worker Bees	Worker bees are female bees; they undertake a multitude of responsibilities within the hive. These duties include nurturing larvae, gathering nectar and pollen, constructing and upkeeping the comb structure, and safeguarding the hive against potential threats from intruders.
Drone Bees	Drones, unlike their female counterparts, are male bees devoid of stingers. Their primary role is mating with queens from different

Bee Type	Description
	colonies. Unlike worker bees, drones do not engage in foraging or any other tasks within the hive.

The Life Cycle of Honeybees

Honeybees undergo an interesting transformation captivating metamorphosis from tiny eggs to fully matured adults, with each stage serving a distinct purpose in the hive ecosystem.

Let's quickly look at the life stages of honeybees:

Egg: Laid by the queen bee in the comb cells, these tiny eggs typically hatch within three days.

Larva: Upon hatching, the larvae emerge and are nourished with a special diet called royal jelly, enabling rapid growth. Larvae rely entirely on worker bees for sustenance.

Pupa: After around six days, the larva encases itself in a cocoon, entering the pupal stage. Within the cocoon, the foundational structures of the adult honeybee take shape.

Adult: Eventually, the fully developed adult bee emerges from the cocoon. These bees are prepared to fulfill their respective roles within the hive, ensuring its continued existence.

Hive Inspection

Hive inspection is a cornerstone of beekeeping, allowing beekeepers to assess the health and condition of their colonies, identify potential issues, and provide necessary care and management.

Regular monitoring of honeybee colonies is essential in beekeeping, with inspection schedules varying based on numerous factors. This article outlines the timing and frequency for conducting beehive inspections, emphasizing key considerations to ensure no crucial inspections are overlooked. The primary objectives of these inspections are to assess the hive's condition, the colony's health, and the accumulation of hive products. Failure to conduct inspections promptly may result in overlooking potential threats to the hive and its inhabitants, leading to detrimental losses for your beekeeping enterprise.

Determining the timing for beehive inspections involves various factors, including both voluntary checks and those prompted by unforeseen events within the hive or apiary.

Scheduled Inspection

Regular beehive inspections, also known as scheduled inspections, are an integral part of beekeeping management practices. These inspections are conducted as part of a planned routine to monitor the progress of your honeybee colony. Typically, beekeepers schedule these inspections every 2-3 weeks to ensure timely identification of any emerging challenges and the effective application of solutions.

Conducting A Scheduled Inspection

Performing a scheduled beehive inspection efficiently requires being both concise and thorough, allowing you to assess numerous beehives within a limited timeframe. During such inspections, it's essential to examine various aspects, including the production and accumulation of beehive products, the presence of the queen bee and her egg-laying activity, the condition of beehive components, and the potential presence of pests, diseases, parasites, and predators.

Additionally, it's crucial to check on any previously administered treatments or feeding provisions and assess their effectiveness. Depending on their condition, treatments and feeders may need to be removed, replenished, or adjusted as necessary.

Scheduled beehive inspections are meticulously planned and can vary in duration, depending on the depth of examination required. It's crucial not to rush through inspections or omit any essential steps, as thoroughness is key to detecting issues early. While it's not necessary to open every part of the hive during each inspection, it's essential to peek into each beehive box and examine a few frames to assess colony health without subjecting them to adverse weather conditions.

Why a Scheduled Beehive Inspections?

Conducting beehive inspections according to a set schedule is deemed the most effective approach. This method allows for thorough preparation, ensuring that inspections are conducted with care and attention to detail. Scheduled inspections afford ample time for preparation and meticulous record-keeping, minimizing the need for rushed assessments.

Moreover, scheduled inspections contribute to enhanced safety for both the beekeeper and the honeybee colony. With sufficient time allocated, beekeepers can implement measures to prevent stings and minimize disruption to the bees during the inspection process. This reduces the risk of harm to both parties involved.

Scheduled inspections are beneficial for beekeepers of all experience levels. Beginners can benefit from the guidance and support of seasoned beekeepers during their initial inspections. This collaborative approach not only instills confidence but also provides an opportunity for beginners to rectify any mistakes and glean valuable insights from experienced mentors.

Special Inspection

There are instances where you may need to conduct an unplanned inspection of your beehive. This could be prompted by the discovery of an issue during a previous inspection or due to urgent concerns about the hive's well-being. When faced with such circumstances, it becomes necessary to address the problem promptly and assess the effectiveness of any interventions implemented.

During an emergency beehive inspection, the primary focus is on addressing the specific issue identified. This allows beekeepers to gauge the success of their interventions and make any necessary adjustments. While attention is primarily directed towards resolving the identified problem, it's essential to conduct a brief assessment of other aspects of hive integrity to ensure comprehensive evaluation and management of any potential issues.

Special beehive inspections, conducted under specific circumstances, are typically swift and concise endeavors. These inspections may need to be repeated several times to monitor the progress of resolving the identified issue. While conducting a series of special inspections, it's essential to assess the effectiveness of the solutions implemented and to remain vigilant for any emerging problems. It's advisable to keep the inspection duration to a minimum to minimize disruption to the honeybee colony. This approach allows the colony to maintain its strength while recovering from the encountered challenges.

Once it's confirmed that the issue has been resolved, and the honeybee colony is secure, any applied solutions, such as feeders, moving screens, entrance reducers, chemical strips, and small hive beetle traps, should be removed from the hive.

Reasons for Special Beehive Inspections

Special beehive inspections may be necessitated by various circumstances within the hive. These include the emergence of a disease outbreak among the bee population, the invasion of pests or parasites, or incidents of predation. Additionally, instances such as the absence or deterioration of the queen bee's health demand immediate attention to prevent adverse effects on the colony. Furthermore, situations like insufficient forage availability may prompt beekeepers to provide supplementary feeding to sustain the colony's health and vitality.

You can avoid Special Beehive Inspections

Preventive measures are paramount to circumvent the necessity for conducting special beehive inspections. Beekeepers who find themselves needing to conduct multiple special inspections throughout a production year often experience diminished yields of beehive products from the affected colonies. Implementing comprehensive apiary management systems and integrated pest and disease control methods is the most reliable approach to prevent the emergence of circumstances and conditions that require special beehive inspections. Furthermore, employing effective solutions to safeguard the apiary against predators and theft alleviates the challenges stemming from their interference.

Timing Considerations for Beehive Inspections

Choosing the right time of day to conduct an apiary inspection is essential for optimal outcomes. Early morning inspections, when honeybees are less active, offer the advantage of requiring less smoke to calm them. This time frame also minimizes the risk of chilling the brood, as the hive gradually warms up with the rising sun. However, initiating an inspection in the morning may provoke aggression in the bees for the remainder of the day, particularly if the hive tends to be aggressive.

Midday inspections present their own set of considerations. At this time, most forager bees are out of the hive, resulting in fewer bees overall to contend with. Additionally, the ample daylight provides excellent visibility for thorough examination. Utilizing sugar water can further aid in calming the bees, as they have sufficient time to groom themselves. Furthermore, the warm temperatures ensure that the hive's humidity returns to normal before nightfall.

Alternatively, some beekeepers opt for evening inspections. If conducted briefly, these inspections can mitigate the risk of chilling brood. Evening inspections capitalize on the bees' tendency to settle in for the night, rendering them less active and aggressive. By the following morning, they typically regain their calm demeanor. However, evening inspections are disadvantaged by poor lighting conditions, and the use of sugar water as a substitute for a bee smoker is not feasible.

Post-Inspection Hive Maintenance

Ensuring the continued well-being of your hive post-inspection is paramount to its longevity and productivity. Here are the necessary steps to maintain your hive effectively:

Secure Frames: Return the frames to their designated positions within the hive body, ensuring they are firmly secured to prevent dislodgment or damage during handling.

Repair Damaged Components: Promptly address any broken parts or equipment to prevent further deterioration and disruption to the colony's activities.

Reinstate Protective Covers: Reattach or replace any protective covers over hive openings removed during inspection. This safeguards against predators and prevents moisture accumulation, which can lead to mold growth.

By diligently following these post-inspection procedures, you can uphold the health and functionality of your hives, mitigating potential issues stemming from neglect or inadequate maintenance practices.

With these tasks completed, you're now prepared to proceed to the crucial task of documenting inspection data – an indispensable aspect of effective beekeeping management!

Recording Inspection Data

After completing the inspection, documenting your observations is crucial for effective hive management. This process serves dual purposes: it aids in recalling and reviewing critical details, while also enabling continuous monitoring of the hive's progression.

To ensure accurate and informative records, consider the following guidelines:

Date and Visual Reference: Begin by recording the date of the inspection. Including a reference photo taken during the inspection can facilitate future comparisons and analysis.

Noting Changes: Document any notable changes or anomalies observed during the inspection, such as the presence of newly constructed queen cells or frames abundant in pollen.

Health and Activity Comments: Make observations about the overall health and activity levels within the colony. This insight provides valuable context regarding environmental fluctuations throughout the year.

Product Harvest Records: Maintain detailed records of all bee products harvested from the hive, storing them securely away from the hive's vicinity. This ensures accurate tracking of production levels over time.

By adhering to these recording practices, beekeepers can develop a comprehensive dataset that offers insights into how external factors influence bee behavior and hive development. With this knowledge, they can implement informed management strategies to optimize hive performance and longevity.

To sum it up, inspecting a beehive stands as a fundamental task in beekeeping, offering insights into hive health, progress, and environmental concerns. To ensure safe and effective inspections, consider the following guidelines:

1. **Preparation is Key:** Before beginning the inspection, ensure your protective gear, including long sleeves, pants, gloves, and a hat with a veil or mesh netting, is in good condition. It's essential to wear proper attire, regardless of your experience level. Check that all clothing fits snugly to prevent bee entry. Additionally, inspect your smoker to ensure it functions

correctly; the smoke aids in keeping bees calm during the inspection.

2. **Start with Care:** Commence the inspection by removing frames from the box, beginning with the outer frames. These frames typically contain honeycomb and stored food supplies rather than delicate eggs or larvae. Handle each frame with care to avoid damaging the comb or injuring bees.

3. **Thorough Assessment:** Examine each frame individually, paying attention to larvae development stages, pollen stores for feeding young bees, and signs of pests or diseases that may threaten the hive. Take note of any changes since the previous inspection and discuss findings with fellow beekeepers for additional insights and recommendations.

4. **Gentle Handling:** Handle frames with care, ensuring they are returned to their original position without shaking off

excess bees, which can cause harm. Maintain a steady and gentle approach throughout the inspection process to minimize disturbances to the hive.

By adhering to these steps during each inspection, you contribute to the safety and well-being of your beehives, fostering healthy hive development and protecting native bee species. With diligent inspections, you can rest assured that your hives thrive in harmony with nature, fulfilling their vital role in pollination and ecosystem preservation.

The frequency of hive inspections depends on various factors, including the time of year, weather conditions, and the developmental stage of the colony. During the active beekeeping season, hive inspections are typically conducted every 7-14 days to monitor colony growth, assess hive health, and address any issues that may arise. In contrast, during the winter months or periods of reduced activity, hive inspections may be less frequent, with beekeepers focusing on providing supplemental feeding and winterizing the hive.

The timing of hive inspections is critical to ensure optimal conditions for both beekeepers and bees. Conduct inspections during the middle of the day when temperatures are warmest and bees are most active. Avoid inspecting hives during extreme weather conditions, such as high winds, heavy rain, or extreme heat, as these conditions can stress the bees and disrupt hive operations.

When conducting hive inspections, employ gentle and methodical techniques to minimize disruption and stress to the bees. Begin by smoking the entrance and top of the hive to calm the bees and reduce defensive behavior. Carefully remove the hive cover and inner cover, taking care to avoid squashing bees or causing unnecessary disturbance. Use a hive tool to gently pry apart hive components, such as frames and boxes, and inspect each frame methodically for signs of brood, honey, pollen, and pests. Take note of any abnormalities or issues observed during the inspection, such as queen cells, swarm preparations, or signs of disease or pest infestation.

Maintain detailed records of hive inspections, including the date, observations, and any actions taken. Recording hive data allows beekeepers to track hive health and performance over time, identify trends or patterns, and make informed management decisions. Consider using a hive inspection checklist or journal to streamline record-keeping and ensure consistency in monitoring hive conditions.

By conducting hive inspections regularly, timing them appropriately, employing gentle techniques, and maintaining detailed records, beekeepers can effectively monitor hive health and performance, identify potential issues, and provide timely care and management to support thriving and productive bee colonies.

Identifying and Managing Common Pests and Diseases

Bee colonies are susceptible to various pests and diseases that can impact their health, longevity, and productivity. Here are some key pests and diseases to be aware of, along with strategies for identification and management:

1. Varroa Mites: Varroa destructor is perhaps the most notorious pest affecting bee colonies worldwide. These parasitic mites feed on the bodily fluids of adult bees and developing brood, weakening the bees and transmitting viruses that can devastate entire colonies. To identify varroa mites, beekeepers can conduct regular mite monitoring using sticky traps or alcohol washes. Management strategies for varroa mites include integrated pest management (IPM) techniques, such as chemical treatments, drone brood removal, and genetic selection for mite-resistant bee stocks.

2. Nosema: Nosema is a fungal parasite that infects the digestive tracts of bees, causing dysentery, reduced lifespan, and weakened immune systems. To identify Nosema infection, beekeepers can conduct microscopic examinations of bee feces for spores. Management strategies for Nosema include sanitation, proper hive ventilation, and supplemental feeding to support colony health and resilience.

3. American Foulbrood (AFB): AFB is a bacterial disease that affects bee brood, causing foul-smelling larvae that turn brown and become ropey in texture. To identify AFB, beekeepers can perform the "rope test," where infected larvae are stretched between two toothpicks to form a rope-like consistency. AFB is highly contagious and can quickly spread throughout a hive, so infected colonies must be promptly destroyed to prevent further transmission to neighboring colonies.

4. Chalkbrood: Chalkbrood is a fungal disease that affects bee brood, causing infected larvae to mummify and turn chalky white in appearance. To identify chalkbrood, beekeepers can inspect brood cells for infected larvae and examine hive debris for chalk-like mummies. Management strategies for chalkbrood include maintaining proper hive hygiene, improving ventilation, and ensuring adequate nutrition for the bees.

5. **Small Hive Beetle (SHB):** Small hive beetles are opportunistic pests that feed on pollen, honey, and bee brood, causing damage to hive infrastructure and contaminating honey stores. To identify SHB infestations, beekeepers can inspect hives for adult beetles, larvae, and slime trails left behind by feeding beetles. Management strategies for SHB include reducing hive entrances, installing beetle traps, and maintaining strong, healthy colonies capable of defending against beetle incursions.

By familiarizing with the signs and symptoms of common pests and diseases, beekeepers can effectively monitor hive health, implement timely interventions, and safeguard their colonies against potential threats. Additionally, proactive hive management practices, such as maintaining proper sanitation, promoting colony resilience, and implementing integrated pest management strategies, can help mitigate the risk of pest and disease outbreaks and support the long-term health and productivity of bee colonies.

Swarm Prevention and Management Strategies

Swarming is a natural behavior exhibited by honeybee colonies as a means of reproduction and expansion. While swarming is a normal part of the bee lifecycle, it can also result in the loss of bees and honey production for beekeepers. Therefore, implementing effective swarm prevention and management strategies is essential for maintaining hive health and maximizing productivity. Here are some key strategies to consider:

1. Provide Adequate Space: One of the primary triggers for swarming is overcrowding within the hive, which can occur when the colony outgrows its available space. To prevent swarming, beekeepers should regularly monitor hive population and expansion and provide additional hive space as needed. This can be accomplished by adding extra hive boxes or supers, allowing the colony room to grow and expand without feeling constrained.

2. Manage Queen Cells: Queen cells are special cells built by worker bees to rear new queen bees, a precursor to swarming. During hive inspections, beekeepers should carefully monitor for the presence of queen cells and take appropriate action to prevent swarming. Depending on the situation, options for managing queen cells may include removing excess queen cells, performing a split to create a new colony, or artificially swarming the colony to preempt a natural swarm.

3. Promote Swarm Readiness: By promoting swarm readiness within the hive, beekeepers can encourage the colony to channel its swarming instincts in controlled ways. This can be achieved through techniques such as swarm management, where beekeepers simulate swarm conditions by creating queenless or queenright splits, providing swarm traps or bait hives, and actively monitoring hive conditions for signs of swarming readiness.

4. Monitor Hive Health: Maintaining optimal hive health is essential for minimizing the likelihood of swarming. Beekeepers should regularly inspect hives for signs of disease, pest infestation, and other stressors that may trigger swarming behavior. By addressing issues promptly and proactively managing hive health, beekeepers can create conditions conducive to hive stability and productivity, reducing the risk of swarming.

5. Provide Adequate Nutrition: Ensuring that bees have access to a diverse and abundant food supply is crucial for maintaining hive strength and preventing swarming. Beekeepers should plant bee-friendly forage crops, provide supplemental feeding when necessary, and monitor hive stores to ensure that bees have an ample supply of nectar and pollen to sustain them during periods of rapid growth and expansion.

Honeybee Nutrition

Feeding bees stands as a vital component of beekeeping, ensuring the sustenance and well-being of bee colonies. Bees rely on a consistent food supply to flourish, and when natural nectar sources are scarce, beekeepers must intervene by providing supplementary nourishment such as sugar water or alternative food options.

This practice becomes especially critical during periods of dearth, such as winter months or in regions with limited floral resources. Without sufficient sustenance, bee colonies face the risk of collapse, posing significant challenges for beekeepers and environmental balance alike. Thus, grasping the fundamentals of bee feeding and exploring diverse feeding methods becomes imperative for every beekeeper's toolkit.

Beekeepers possess a range of feeding alternatives, encompassing sugar water, pollen patties, and fondant. Each approach carries its own set of merits and drawbacks, with the selection of a feeding method contingent upon the beekeeper's objectives and the requirements of their colonies. Through furnishing bees with essential nutrients, beekeepers play a pivotal role in fostering the well-being and vigor of their colonies. This proactive approach not only enhances honey production but also fosters a more sustainable framework for beekeeping operations.

Macronutrients

Macronutrients are vital for bees and are needed in significant amounts. The primary essential nutrients required by bees include carbohydrates, proteins, and lipids. Carbohydrates, sourced from nectar and honey, serve as the main energy source for bees. Proteins are derived from pollen, serving as the primary reservoir of amino acids crucial for bee health. Lipids, sourced from both nectar and pollen, play a key role in the synthesis of beeswax and other essential bee products.

Micronutrients

Essential nutrients are vital for bees and are required in small quantities, known as micronutrients. These encompass vitamins, minerals, and other trace elements crucial for maintaining bee health and bolstering their immune system. Calcium, for instance, plays a pivotal role in larval development, while magnesium is indispensable for wax production.

Micronutrients are sourced from various outlets, predominantly pollen and nectar. Pollen, abundant in micronutrients like vitamins and minerals, serves as a rich source. Notably, different pollen types offer distinct nutritional compositions, prompting bees to select specific varieties based on their dietary requirements.

Basics of Bee Feeding

Feeding bees is a critical component of successful beekeeping, ensuring their sustenance and prosperity. This section outlines the fundamentals of bee feeding, encompassing the timing and constituents of bee nutrition.

Beekeepers must provide supplemental food during periods of natural food scarcity, which may arise during droughts, frigid weather, or when floral resources are limited. Typically, bee feeding is crucial in autumn and winter to fortify colonies for the colder months.

Nutritional Sources

Sugar syrup stands as the primary sustenance for bees, created by dissolving sugar in water. Alternatively, beekeepers can offer natural options like honey and pollen. Honey boasts essential nutrients and minerals, while pollen serves as the principal protein source for bees. Access to water is equally vital for bees, aiding in hive temperature regulation and honey dilution. Beekeepers can facilitate water access by situating shallow water dishes near hives.

Considerations for Climate

Climate considerations are pivotal in bee feeding practices. Bees' nutritional requirements vary with climate, necessitating adjustments accordingly. In colder climates, bees require larger food stores to endure the winter months.

Varieties of Bee Feeding Equipment

Bee feeding equipment encompasses a range of types and designs, each with distinct functionalities and characteristics. Among the most prevalent feeders are frame feeders, hive top feeders, boardman feeders, and entrance feeders. The selection of feeder depends on the beekeeper's preferences and the specific needs of the colony.

Frame Feeders

These feeders are positioned within the hive, fitting snugly between two frames. Typically crafted from plastic or wood, they have the capacity to contain up to one gallon of syrup. Frame feeders facilitate easy refilling without necessitating hive opening. However, they carry the risk of bee drowning.

Hive Top Feeders

Situated atop the hive, hive top feeders are adept at holding substantial syrup volumes. Usually constructed from plastic or metal, they can accommodate up to five gallons of syrup. Similar to frame feeders, they offer convenient refill options without requiring hive access. Nevertheless, they tend to be costly and may contribute to moisture accumulation within the hive.

Boardman Feeders

Situated at the hive entrance, boardman feeders are crafted to contain a modest quantity of syrup. Typically fashioned from plastic or glass, they have the capacity to hold up to one quart of syrup. These feeders offer easy replenishment without necessitating hive intrusion. However, they are susceptible to bee robbing and are not recommended for winter use.

Placement and Maintenance of Feeders

When situating feeders, careful attention must be paid to both their placement and capacity. Placing the feeder too close to the hive entrance may attract unwanted attention from other bees, leading to robbing. Conversely, positioning it too far from the colony may result in the bees failing to locate it.

Regular maintenance of feeders is essential to ensure a continuous syrup supply for the bees. Cleaning the feeder at regular intervals prevents mold and bacteria growth. Additionally, inspecting the feeder for leaks and cracks helps prevent syrup spillage within the hive.

Seasonal Feeding

Beekeepers must remain attuned to the seasonal fluctuations and nutritional requirements of their colonies throughout the year. Here are various seasonal feeding tactics to support the vitality and endurance of honeybee populations.

Spring Feeding

Spring marks a pivotal period for bee colonies, as they strive to replenish their numbers and amass sufficient food reserves for the impending season. Beekeepers should diligently oversee their hives and administer supplementary feeding with sugar syrup or fondant as needed.

Summer Management

In the summer months, bees engage in active foraging to gather and store honey, vital for sustaining the colony through the upcoming fall and winter seasons. Beekeepers play a crucial role in ensuring the hive has ample space to accommodate the incoming honey and pollen. Moreover, if the colony has recently swarmed or undergone a split, beekeepers can anticipate the emergence of mated queens initiating egg-laying activities. For many beekeepers, this period heralds the time for honey harvesting.

Fall Preparations

As autumn approaches, bees diligently prepare for the rigors of winter. Beekeepers should diligently monitor hive stores, ensuring an adequate honey reservoir for the colder months ahead. In cases where stores are insufficient, supplementation with sugar syrup or dry sugar becomes imperative to bolster the hive's survival prospects. Insufficient food reserves render hives incapable of generating sufficient heat to combat the cold, jeopardizing their chances of survival. Additionally, beekeepers should remain vigilant against the threat of mites and other pests, which can undermine the colony's strength and resilience.

Winter Survival

The winter months present the greatest test for bee colonies, as they rely solely on their accumulated food reserves for sustenance. Beekeepers play a crucial role in ensuring that hives are adequately stocked with honey and pollen to endure the cold season. In cases of insufficient stores, beekeepers may resort to supplemental feeding, utilizing options like fondant or dry sugar. However, it's essential to exercise caution during winter feeding, as opening the hive risks dissipating precious heat and energy vital for the bees' survival. Seeking guidance from seasoned beekeepers or beekeeping organizations can provide valuable insights into effective winter feeding strategies.

Nourishing Colony Health

Ensuring the proper nourishment of bees stands as a fundamental aspect of successful beekeeping. A colony that receives ample sustenance demonstrates enhanced resilience against diseases and pests. Moreover, a well-fed colony exhibits improved brood production, a vital factor contributing to colony survival and expansion.

Observing and Modifying Feeding Practices

Beekeepers have a crucial responsibility to regularly assess their hives to guarantee an adequate food supply for their bees. During these inspections, it's imperative for beekeepers to evaluate the honey reserves to ascertain if they meet the bees' requirements. In cases where the honey stores are insufficient, beekeepers must take action to provide supplemental feeding.

Furthermore, beekeepers should remain vigilant about temperature fluctuations and weather patterns to determine if additional feeding is necessary. Cold weather, in particular, demands extra attention as bees require more sustenance to endure the winter months. Hence, it's paramount for beekeepers to ensure that their colonies possess ample food reserves to withstand the cold.

However, if the bees already possess adequate food stores, beekeepers should refrain from feeding them unnecessarily. Overfeeding can foster dependency on the beekeeper, potentially weakening the colony in the long run. Thus, prudent monitoring and adjustment of feeding practices are essential to maintain the health and independence of bee colonies.

Best Bee Feeding Practices

Feeding bees demands adherence to best practices to guarantee their nutritional requirements are met effectively. Here are recommended strategies for feeding bees:

Select an appropriate feeder tailored to the hive's needs Various feeder types are available, such as top feeders, entrance feeders, and frame feeders.

Employ the correct food source Beekeepers should utilize a sugar syrup concocted from granulated sugar and water. The ideal sugar-to-water ratio is 2:1 for fall feeding and 1:1 for spring feeding.

Avoid feeding bees during a nectar flow period Bees naturally forage nectar from flowers when it's accessible, hence beekeepers should refrain from supplemental feeding during such periods.

Ensure access to water Bees rely on water for food digestion, emphasizing the necessity for beekeepers to furnish a nearby water source for their colonies.

Mitigating Disease and Pest Risks

Maintaining a balanced diet for bees plays a pivotal role in averting diseases and pest invasions. Inadequate nutrition compromises the bees' immune system, rendering them more vulnerable to infections and attacks. For instance, insufficient protein intake can lead to subpar brood production, thereby weakening the colony's defense against pests like the small hive beetle.

To shield colonies from diseases and pests, beekeepers must ensure they receive a balanced diet comprising carbohydrates, proteins, vitamins, and minerals. Additionally, regular monitoring of colonies for signs of ailments and infestations is essential, enabling beekeepers to intervene promptly when necessary.

Summarily, Bees rely on a diverse and balanced diet of nectar and pollen to meet their nutritional needs throughout their lifecycle. As responsible stewards of bee colonies, beekeepers play a vital role in ensuring that bees have access to the nutrition they need to thrive. Here are some key considerations for providing adequate nutrition for bees:

1. Forage Diversity: Bees require a diverse range of floral resources to meet their nutritional requirements fully. Beekeepers should aim to provide bees with access to a variety of flowering plants, trees, and shrubs that offer abundant sources of nectar and pollen throughout the seasons. Planting a diverse array of bee-friendly forage crops, such as wildflowers, fruit trees, and herbs, can help ensure that bees have access to a balanced diet and essential nutrients year-round.

2. Seasonal Considerations: Beekeepers should be mindful of the seasonal fluctuations in nectar and pollen availability and take proactive measures to supplement bees' diets during periods of scarcity. In early spring, when natural forage is limited, bees may require supplemental feeding to stimulate brood rearing and colony growth. Similarly, in late summer and fall, when floral resources are scarce, providing bees with supplemental feeding can help ensure that colonies have sufficient stores to sustain them through the winter months.

3. Supplemental Feeding: Supplemental feeding involves providing bees with additional sources of nectar and pollen to supplement their natural forage. Beekeepers can use various feeding methods, such as sugar syrup, pollen patties, and protein supplements, to provide bees with the nutrients they need to thrive. When supplementing bees' diets, it's essential to use high-quality, bee-appropriate food sources and monitor hive conditions closely to prevent overfeeding or resource competition.

4. **Water Sources:** In addition to nectar and pollen, bees also require access to clean, fresh water for hydration and hive cooling. Beekeepers should provide bees with access to a reliable water source, such as a shallow pond, birdbath, or dripping faucet, located within close proximity to the hive. Ensure that water sources are easily accessible and free from contaminants to support bee health and vitality.

5. **Hive Health Monitoring:** Regular hive inspections and monitoring are essential for assessing hive health, identifying nutritional deficiencies, and implementing appropriate interventions. Beekeepers should closely observe hive conditions, including brood patterns, honey stores, and bee behavior, to identify signs of

malnutrition or stress and take proactive steps to address any issues that may arise.

By prioritizing forage diversity, seasonal considerations, supplemental feeding, water provision, and hive health monitoring, beekeepers can ensure that their bees have access to the nutrition they need to thrive, promoting strong, healthy colonies capable of withstanding environmental challenges and maximizing productivity.

CHAPTER FIVE
UNDERSTANDING BEE LIFE CYCLE AND COLONY DYNAMICS

The life cycle of a honeybee encompasses various stages, each with its unique characteristics and roles within the colony. Understanding these stages and their dynamics is essential for beekeepers to effectively manage and care for their bee colonies, starting from the egg to adult stage as described below:

1. Egg Stage: The life cycle of a honeybee begins when the queen bee lays an egg in a wax cell within the hive. Fertilized eggs develop into female worker bees, while unfertilized eggs develop into male drones. The queen bee controls the sex of the eggs by selectively fertilizing them with sperm stored in her spermatheca.

2. Larval Stage: After hatching from the egg, the bee larva is fed a diet of royal jelly by nurse bees for the first few days of its life. As the larva grows, it is gradually weaned onto a diet of pollen and honey. Larvae undergo several molts as they grow and develop, eventually pupating within their wax cells.

3. Pupal Stage: During the pupal stage, the larva undergoes metamorphosis, transforming into an adult bee. Inside the wax cell, the pupa undergoes dramatic changes, including the development of wings, legs, eyes, and other body structures. The pupal stage lasts for approximately 12 days for worker bees, 14 days for drones, and 16 days for queens.

4. Adult Stage: Once the pupal stage is complete, the adult bee chews its way out of its wax cell and emerges into the hive as a fully developed bee. Depending on their role within the colony, adult bees assume various tasks and responsibilities, including nursing brood, foraging for food, guarding the hive, and tending to the queen.

Colony dynamics within a honeybee colony are governed by complex social interactions and division of labor among colony members. The colony's population consists of three primary castes: the queen, workers, and drones, each with its specific roles and responsibilities. The queen bee is responsible for laying eggs and maintaining the genetic diversity of the colony, while worker bees perform various tasks, such as brood care, foraging, and hive maintenance. Drones, on the other hand, serve a reproductive role, mating with virgin queens from other colonies.

Throughout the year, honeybee colonies undergo dynamic changes in population size, behavior, and resource utilization in response to environmental factors, seasonal fluctuations, and internal colony dynamics. Beekeepers must closely monitor colony dynamics, population trends, and behavior to ensure the health, stability, and productivity of their colonies.

By understanding the intricacies of bee life cycles and colony dynamics, beekeepers can effectively manage and care for their bee colonies, promoting strong, resilient populations capable of thriving in their environment and maximizing honey production.

Roles Within the Hive

Let's explore the fascinating roles played by the three primary castes within a honeybee hive: the queen, workers, and drones. Each caste has distinct responsibilities and contributions that collectively ensure the health, productivity, and survival of the colony.

1. Queen Bee:

  - The queen bee is the mother of the colony, responsible for laying eggs and maintaining the genetic diversity of the hive.

  - She is the largest bee in the colony, with a long abdomen and distinctive physiology adapted for egg-laying.

  - The queen's primary role is to lay fertilized eggs, which develop into worker bees and new queen bees, ensuring the colony's continued growth and vitality.

  - In addition to egg-laying, the queen produces pheromones that regulate colony behavior, suppress worker reproduction, and maintain colony cohesion.

- The queen bee is tended to and cared for by worker bees, who feed her, groom her, and ensure her comfort and well-being.

2. Worker Bees:

- Worker bees are the backbone of the hive, comprising the vast majority of the colony's population and performing a wide range of essential tasks to support colony function and survival.

- Workers are female bees that do not mate and are sterile, meaning they cannot lay fertilized eggs.

- Throughout their lifespan, worker bees assume various roles within the hive, including nursing brood, foraging for food, guarding the hive entrance, building comb, and maintaining hive cleanliness.

- Worker bees transition through a series of temporal tasks, beginning as nurse bees responsible for caring for the queen and brood, then transitioning to roles such as pollen or nectar collectors, and eventually serving as guard bees or undertakers responsible for removing dead bees and debris from the hive.

- The collective efforts of worker bees ensure the day-to-day functioning of the hive, from rearing new generations of bees to storing food, regulating hive temperature, and defending against threats.

3. Drone Bees:

- Drones are male bees produced by unfertilized eggs laid by the queen bee.

- Drones are larger and bulkier than worker bees, with distinctive large eyes and a stout body adapted for flight and mating.

- The primary role of drones is to mate with virgin queens from other colonies, contributing to the genetic diversity of the bee population.

- Drones do not engage in typical hive tasks such as foraging or nursing brood; instead, they rely on worker bees for food and care.

- Drones typically emerge in greater numbers during the spring and summer months when mating flights occur, and are expelled from the hive in the fall when resources become scarce.

Brood Development and Hive Population Dynamics

Brood development refers to the lifecycle stages of honeybee brood, encompassing the egg, larval, and pupal stages, culminating in the emergence of adult bees. This process is closely intertwined with hive population dynamics, which involve fluctuations in the number and composition of bees within the colony. Now, let us break it down for proper understanding:

1. **Egg Stage:** The brood development cycle begins when the queen bee lays eggs in wax cells within the hive. Fertilized eggs develop into female worker bees, while unfertilized eggs develop into male drones. The queen controls the sex of the eggs by selectively fertilizing them with sperm stored in her spermatheca.

2. **Larval Stage:** After hatching from the egg, the bee larva is fed a diet of royal jelly by nurse bees for the first few days of its life. As the larva grows, it is gradually weaned onto a diet of pollen and

honey. Larvae undergo several molts as they grow and develop, eventually pupating within their wax cells.

3. Pupal Stage: During the pupal stage, the larva undergoes metamorphosis, transforming into an adult bee. Inside the wax cell, the pupa undergoes dramatic changes, including the development of wings, legs, eyes, and other body structures. The pupal stage lasts for approximately 12 days for worker bees, 14 days for drones, and 16 days for queens.

4. Adult Stage: Once the pupal stage is complete, the adult bee chews its way out of its wax cell and emerges into the hive as a fully developed bee. Adult bees assume various roles and responsibilities within the colony, including nursing brood, foraging for food, guarding the hive, and tending to the queen.

Hive population dynamics are influenced by factors such as seasonality, resource availability, environmental conditions, and colony health. During the active beekeeping season, hive populations typically experience periods of growth and expansion as new brood emerges and matures, leading to an increase in the number of worker bees within the colony. Conversely, during the winter months or periods of reduced activity, hive populations may decline as bees die off or enter a state of dormancy to conserve energy and resources.

Beekeepers play a crucial role in managing hive population dynamics to ensure the health, stability, and productivity of their colonies. By monitoring brood development, assessing hive conditions, and implementing appropriate management practices, beekeepers can support optimal hive population growth, minimize swarming, and promote colony resilience in the face of environmental challenges.

Seasonal Considerations and Hive Management

Beekeepers must consider the seasonal variations in weather, forage availability, and colony dynamics when managing their hives. Here's a breakdown of seasonal considerations and their implications for hive management:

1. Spring:

- Spring is a period of renewal and growth for honeybee colonies, marked by increasing temperatures, longer daylight hours, and abundant floral resources.

- During spring, beekeepers focus on stimulating colony growth, brood rearing, and honey production by providing supplemental feeding, monitoring hive health, and conducting regular hive inspections.

- Spring is also a critical time for swarm prevention and management, as colonies may become overcrowded and prone to swarming as populations increase and forage resources become more abundant.

2. Summer:

- Summer is a period of peak activity for honeybee colonies, characterized by warm temperatures, high forage availability, and intensive foraging activity.

- During summer, beekeepers focus on maintaining hive health, managing pests and diseases, and maximizing honey production by providing adequate ventilation, managing varroa mite infestations, and harvesting surplus honey as needed.

- Beekeepers may also take measures to prevent overheating and dehydration in hives by providing shade, supplemental water sources, and adequate ventilation to keep colonies cool and hydrated during hot weather.

3. Fall:

- Fall marks the transition from the active beekeeping season to the onset of winter dormancy, characterized by declining temperatures, shorter daylight hours, and diminishing forage availability.

- During fall, beekeepers focus on preparing colonies for winter by assessing hive health, reducing hive entrances to prevent robbing, and ensuring that colonies have sufficient stores of honey and pollen to sustain them through the winter months.

- Fall is also a time for pest and disease management, as colonies may be more vulnerable to infestations and infections due to declining populations and environmental stressors.

4. Winter:

- Winter is a period of dormancy and survival for honeybee colonies, characterized by cold temperatures, limited forage availability, and reduced colony activity.

- During winter, beekeepers focus on protecting colonies from cold weather, wind, and moisture by insulating hives, providing supplemental feeding when necessary, and monitoring hive conditions for signs of stress or disease.

- Beekeepers may also take measures to mitigate the risk of moisture buildup in hives, such as installing hive insulation, providing upper ventilation, and using moisture-absorbing materials to prevent condensation.

By understanding the seasonal considerations and adapting their management practices accordingly, beekeepers can support the health, productivity, and survival of their honeybee colonies throughout the year. Effective hive management requires careful observation, proactive intervention, and a deep understanding of the unique needs and behaviors of honeybees in different seasons.

CHAPTER SIX
HARVESTING HONEY AND OTHER HIVE PRODUCTS

Assessing Honey Flow and Readiness for Harvest

Honey flow refers to the period when nectar-producing plants are in bloom and bees are actively foraging and collecting nectar to produce honey. Assessing honey flow involves monitoring the availability of nectar sources, observing foraging activity, and evaluating hive conditions to determine when honey is ready for harvest. The below step-by-step guide will assist in assessing honey flow and readiness for harvest:

1. Monitor Foraging Activity: Observing the foraging activity of bees is a key indicator of honey flow. During periods of abundant nectar availability, bees will exhibit increased foraging activity, with workers returning to the hive laden with pollen and nectar. Pay attention to the number of bees coming and going from the hive, the types of flowers they are visiting, and the quantity of pollen and nectar being brought back to the hive.

2. Inspect Hive Stores: Regularly inspecting hive stores is essential for assessing honey flow and readiness for harvest. Check honey supers and brood chambers for signs of honey accumulation, such as capped honey cells, full frames, and visible honey bands. Use a hive tool to gently lift frames and assess their weight and appearance. Frames containing fully capped honey are typically ready for harvest, while frames with uncapped or partially capped honey may require additional time for ripening.

3. Conduct Hive Inspections: During hive inspections, carefully assess hive conditions, brood patterns, and overall colony health to determine the impact of honey flow on colony dynamics. Look for signs of overcrowding, swarm preparations, or congestion in the brood nest, which may indicate that the colony is experiencing a honey flow and requires additional space or management interventions to prevent swarming.

4. Use Visual and Tactile Cues: Visual and tactile cues can also help assess honey flow and readiness for harvest. Inspect frames for signs of ripening honey, such as capped cells, glistening surfaces, and a thick, viscous consistency. Test the ripeness of honey by gently tilting frames and observing how honey flows or drips from capped cells. Fully ripened honey will flow slowly and evenly, with a smooth, glossy appearance.

5. Consider Environmental Factors: Environmental factors such as weather, temperature, and floral availability can influence honey flow and readiness for harvest. Monitor weather forecasts and seasonal patterns to anticipate changes in nectar flow and foraging conditions. Be prepared to adjust hive management practices accordingly, such as adding or removing honey supers, providing supplemental feeding, or protecting hives from adverse weather conditions.

By carefully assessing honey flow and readiness for harvest, beekeepers can maximize honey production, ensure the quality of their honey crop, and support the health and productivity of their honeybee colonies. Effective hive management requires keen observation, attention to detail, and a deep understanding of the unique dynamics of honey flow in different seasons and environments.

Harvesting Methods

Let us examine the various methods of harvesting honey—extracting, crushing, and straining—each offering unique advantages and considerations for beekeepers looking to harvest their honey crop.

1. Extracting:

   - Extracting honey is the most common method used by commercial beekeepers and hobbyists alike. It involves removing honey frames from the hive, uncapping the cells using a hot knife or uncapping fork, and then placing the frames in a honey extractor—a centrifugal machine that spins the frames, extracting honey through centrifugal force.

- Extracting is preferred for harvesting large quantities of honey efficiently and preserving the integrity of the honeycomb. It allows beekeepers to extract honey from frames without damaging the comb, enabling bees to reuse the comb for future honey storage.

- After extraction, honey is typically filtered to remove any remaining wax or debris before being bottled and sealed for storage or sale.

2. Crushing and Straining:

- Crushing and straining is a simpler method of honey harvesting that involves crushing honeycomb to release the honey, then straining the honey through a mesh or cheesecloth to remove wax and debris.

- This method is ideal for beekeepers with small-scale operations or those who prefer a more hands-on approach to honey harvesting. It requires minimal equipment and is suitable for beekeepers who do not have access to honey extractors or prefer not to use them.

- While crushing and straining may result in a slightly lower yield compared to extraction, it preserves the natural structure of the honeycomb and allows beekeepers to harvest honey without specialized equipment.

3. Considerations:

- When harvesting honey, beekeepers should ensure that honey is harvested at the appropriate moisture content (around 18-20%) to prevent fermentation and spoilage.

- It's essential to handle honey with care to preserve its quality and flavor. Avoid exposing honey to excessive heat or agitation, as this can cause honey to lose its natural aroma, flavor, and nutritional properties.

- Proper sanitation and hygiene are crucial during honey harvesting to prevent contamination and ensure food safety. Use clean, food-grade equipment and containers, and follow good hygiene practices to maintain the quality and purity of the honey.

Processing Other Hive Products
1. Beeswax:

- Beeswax is a natural wax produced by honeybees and is used by bees to construct honeycomb cells for storing honey and raising brood.

- Beekeepers can harvest beeswax by removing old or excess comb from the hive and melting it down to extract the wax.

- Once harvested, beeswax can be processed and purified to remove impurities and debris, resulting in a clean, golden-yellow wax suitable for various applications.

- Beeswax has numerous uses, including candle making, skincare products, cosmetics, woodworking, and art and craft projects. It is prized for its natural fragrance, versatility, and skin-nourishing properties.

2. Propolis:

- Propolis, also known as bee glue, is a resinous substance collected by honeybees from tree buds, sap flows, and other botanical sources. Bees use propolis to seal cracks and gaps in the hive, reinforce hive structure, and protect against pathogens and pests.

- Beekeepers can collect propolis by scraping it off hive surfaces or using propolis traps designed to collect excess propolis.

- Once collected, propolis can be processed and purified to remove wax and debris, resulting in a concentrated resin with antimicrobial, antioxidant, and anti-inflammatory properties.

- Propolis has a wide range of applications in natural medicine, skincare products, oral health products, and dietary supplements. It is valued for its immune-boosting properties and potential health benefits.

3. Royal Jelly:

- Royal jelly is a nutrient-rich secretion produced by worker bees and fed to queen larvae during their development. It is rich in proteins, vitamins, minerals, and fatty acids, making it an essential food source for queen bees and larvae.

- Beekeepers can harvest royal jelly by grafting queen cells and collecting the royal jelly produced by nurse bees.

- Once harvested, royal jelly can be processed and preserved through freezing or lyophilization (freeze-drying) to maintain its nutritional integrity.

- Royal jelly is highly valued for its potential health benefits and is used in dietary supplements, skincare products, and traditional medicine. It is believed to have anti-aging, immune-boosting, and fertility-enhancing properties.

By processing and utilizing these hive products—beeswax, propolis, and royal jelly—beekeepers can maximize the value of their beekeeping operations while promoting sustainability, health, and well-being. These natural products offer a wealth of opportunities for personal use, commercial ventures, and creative endeavors, contributing to a thriving and diverse beekeeping industry.

CONCLUSION

In conclusion, "The Backyard Beekeeper's guide" provides a comprehensive exploration of the fascinating world of beekeeping, offering beekeepers of all levels invaluable insights, practical tips, and sustainable practices for successfully managing honeybee colonies and promoting pollinator health. Throughout the book, we have delved into various aspects of beekeeping, from understanding bee anatomy and behavior to addressing common hive issues, navigating environmental challenges, and embracing innovations in beekeeping technology.

As beekeepers, we are stewards of these remarkable creatures, entrusted with the responsibility of nurturing and safeguarding honeybee colonies for the benefit of ecosystems, agriculture, and human well-being. By adopting sustainable beekeeping practices and advocating for pollinator conservation, we can make a positive impact on the health and resilience of honeybee populations and the broader environment.

The challenges facing beekeepers—climate change, pesticides, varroa mites—are formidable, but they are not insurmountable. Through collective action, collaboration, and a commitment to continuous learning and adaptation, beekeepers can overcome these challenges and thrive in the pursuit of sustainable beekeeping.

The Backyard Beekeeper's guide is not just a book—it is a testament to the enduring bond between humans and bees, a celebration of the wonders of nature, and a call to action for beekeepers everywhere to embrace their role as guardians of the hive. We hope this guide serve as a source of knowledge, inspiration, and empowerment for beekeepers of all backgrounds, as we work together to cultivate thriving honeybee colonies and a more sustainable future for generations to come.

www.ingramcontent.com/pod-product-compliance
Lightning Source LLC
Chambersburg PA
CBHW070250230526
45470CB00002B/550